The Unified Theory of Probability

Introducing the Concept of Probability Error

I0490649

BY

ELIPHAS PHIRI

MttC Publications

Message to the Christian Publications

Copyright © 2024 2ND ED

Message to the Christian Publications

All rights reserved.

ISBN:9798388863737
EMAIL: phirieliphas2@gmail.com

DEDICATION

To all students of Mathematics and Physics

Contents

Part 1

The Unified Theory of Probability

Chapter 1

Probability Errors

Factors Affecting the Probability of Events

There is nothing like pure chance. All probability events are affected by certain

11

unwanted factors analogous to errors in science. We can call these errors Probability errors.

The factors that can affect the probability can be systematic or random just as in science.

1) Systematic Errors

This is the error that affect the probability of an event due to the defects of the objects involved. For instance, a coin may have a defect so that its probability is affected by that defect.

2) Random Errors

This is due to factors that vary. This error may be due to human error e.g. the way someone tosses the coin, or environmental factors e.g. the wind blowing on a coin tossed in the air.

3) Combined Errors

The error due to both systematic and random errors.

Theoretical Probability

Theoretical Probability is given by the formula;

Probability of Event = number of ways the event can happen /total number of events that can happen

i.e. $P_{Th} = \dfrac{n(E)}{n(S)}$

This probability is ideally never affected by experimental errors.

Experimental Probability

The probability which is found as a result of an experiment/trial is called Experimental Probability. The experiment maybe performed by humans or by nature i.e.

Probability of Event = number of ways the event happened /total number of events happened (total trials) i.e. $P_{EX} = \dfrac{n(E)}{n(S)}$

This is the probability that is always affected by Probability Errors.

Mean Experimental Probability

The definition of experimental probability is not complete. The number of trials is supposed to be stated to show that the probability was found after a certain number of trials. The experimental probabilities will not be the same for different number of trials e.g. the probability at 100 trials will not be the same as the probability at 1000 trials. We would expect the experimental probability to approach the theoretical probability as the number of trials increase.

The average of the experimental probability at T number of trials must also be found. Doing one trial and calling it the probability of that event is not sufficient.

So we should be talking about the average of experimental probability.

So I Suggest experimental Probability of Event should be defined as follows, Probability = number of outcomes of the event /number of trials performed, at number of trials, T

i.e. $P_{Exp} = \dfrac{n(E)}{n(T)}$, at number of trials = T

Mean experimental probability = sum of number of experimental probabilities at various Ts / number of sets of trials(s)

i.e. $\bar{P}_{Exp} = \sum \dfrac{n(C)}{n(T)}/S$

at number of trials = sum of Ts, S = Number of sets of probabilities

---

For example suppose the probabilities of head after tossing a coin ten times in five sets are as shown;

5/10
4/10
6/10
5/10
3/10

Experimental Probability Head = 5/10 or 1/2 for the first trial at T = 10, 4/10 or 2/5 for the second trial at T = 10 and so on.

Then the Mean experimental probability of head is sum of all probabilities divided by 5 i.e. 23/10 divided by 5 = 2.3/5 = 0.46, at T = 10 +10+10+10+10=10 X 5 =50

And theoretical probability of head =0.5.

Chapter 2

Defining Probability Error

Probability Error Formula

The difference between the Theoretical Probability and the mean Experimental Probability shows the probability of the error that affect experimental probability. That is;

Probability Error = Theoretical Probability – Mean Experimental Probability at T

i.e. $P_E = P_{Th} - \bar{P}_n = \dfrac{n(E)}{n(S)} - \dfrac{\sum n(C)}{n(T)s}$, at T trials

---

This error can be defined as the fraction that must be added to experimental probability to get the theoretical probability.

This formula unifies the Experimental Probability and the Theoretical Probability.

This error can be either be positive or negative. For instance, suppose a coin is

17

tossed 10 times and heads comes up 4 times then the experimental probability is 2/5. The theoretical probability is 1/2. So Probability errors for heads = ½ − 2/5 = 1/10.

Then Probability error for tails = ½ − 3/5 = -1/10.

Types of Probability Errors

The probability errors can be systematic or random or the combination of the two analogous to experimental errors in science.

Systematic probability error will therefore be due to the nature of the event or trial. These errors are out of the control of the human.

While the random errors are at the control of the human. So the random probability error can approach the theoretical probability since the factors or errors that affect it are at the messy of the human and so can be eradicated or reduced by performing many trials.

The systematic probability error on the other hand will never approach the theoretical probability since the error is due to the nature of the event or material defect in the event e.g. a biased die. The combined probability error will also never approach the theoretical probability since the systematic errors cannot be eradicated.

INFINITE OR REAL PROBABLITY ERRORS

The difference between the theoretical experimental probability (P_{th}) and the mean experimental probability at T = infinitly many trials (P_∞) can be called the real experimental probability error (RP_E).

These can be divided into three parts according to the three types of errors i.e. systematic probability (P_{ES}), random probability error (P_{ER}) and combined probability error (P_{EC}). This probability can never be found since the experimental probability at T = infinity can never be found.

$P_{ES} = P_{th} - P_\infty$	SYSTEMATIC

$P_{ER} = P_{th} - P_\infty$	RANDOM
$P_{EC} = P_{th} - P_\infty$	COMBINED

FINITE OR PRACTICAL THEORETICAL PROBABLITY ERRORS

The difference between the theoretical experimental probability (P_{th}) and the mean experimental probability at T = m trials (P_m) can be called the practical experimental probability error (PP_E).

These can be divided into three parts according to the three types of errors i.e. systematic practical probability (PP_{ES}), random probability error (PP_{ER}) and combined probability error (PP_{EC}).

The practical systematic and combined errors can never approach zero since the experimental systematic probability and the experimental combined error probability can never approach the theoretical probability. But the practical random probability error can approach zero since the experimental random probability P_m can approach the theoretical probability.

$PP_{ES} = P_{th} - \bar{P}_m$	SYSTEMATIC
$PP_{ER} = P_{th} - \bar{P}_m$	RANDOM
$PP_{EC} = P_{th} - \bar{P}_m$	COMBINED

Chapter 3

Defining Experimental Probability Error

Experimental Probability Error

The difference between the mean experimental probability at $T = n$ and the mean experimental probability at $T = m$ can be called the experimental probability error(XP_E).

21

Experimental Probability Error Types

These errors can be of three types

> 1) Systematic experimental probability error can be expressed as the mean systematic experimental probability at n – systematic experimental probability at infinite.

$$\text{i.e. } XP_{ES} = P_n - P_\infty$$

This error can never approach random error or theoretical error because it is due to defects of the objects involved. It also cannot be known because the probability at infinite cannot be calculated for discrete probability.

> 2) Random probability error can be expressed as the mean random experimental probability at n – random experimental probability at infinite. But random experimental probability error at infinite can approach theoretical probability error for random errors as the number of human and

environmental factors reduce through many trials.

$$\text{So } XP_{ER} = \bar{P}_n - P_\infty = \bar{P}_n - P_{Th}$$

3) combined probability error i.e. combined probability error = mean random experimental probability at n – random experimental probability at infinite

$$\boxed{XP_{EC} = \bar{P}_n} - P_\infty$$

This error cannot be found for sure because of systematic errors which cannot be eliminated by having many number of trials.

$XP_{ES} = \bar{P}_n - P_\infty$	SYSTEMATIC
$XP_{ER} = \bar{P}_n - P_\infty$	RANDOM
$XP_{EC} = \bar{P}_n - P_\infty$	COMBINED

FINITE OR PRACTICAL EXPERIMENTAL PROBABLITY ERRORS

The difference between the mean experimental probability at m (P_m) and the mean experimental probability at T = n trials ($\bar{P}_n$) can be called the practical experimental probability error (PXP_E).

These can be divided into three parts according to the three types of errors i.e. systematic probability (PXP_{ES}), random probability error (PXP_{ER}) and combined probability error (PXP_{EC}).

$PXP_{ES} = \bar{P}_m - \bar{P}_n$	SYSTEMATIC
$PXP_{ER} = \bar{P}_m - \bar{P}_n$	RANDOM
$PXP_{EC} = \bar{P}_m - \bar{P}_n$	COMBINED

CHAPTER 4

PROBABILTY ERROR THEORY

Probability Error of Independent Events

If A and B are independent Events then the probability error is;

$$P_E \text{ (A and B)} = P_E(A)P_E(B)$$

$$= [P_{Th}(A) - P_{Exp}(A)] \, [P_{Th}(B) - P_{Exp}(B)]$$

$$P_{Fct}(A \text{ and } B) = P_{Th}(A)P_{Th}(B) - P_{Exp}(A)P_{Th}(B) - P_{Exp}(B)P_{Th}(A) + P_{Exp}(A)P_{Exp}(B)$$

Probability Error of dependent Events

If A and B are dependent Events then the probability error is;

$$P_E \text{ (A and B)} = P_E(A)P_E(B \text{ following A})$$

$$= [P_{Th}(A) - P_{Exp}(A)] \, [P_{Th}(B \text{following A}) - P_{Exp}(B \text{following A})]$$

P_E (A and B) $= P_{Th}(A)P_{Th}(Bfollowing A) - P_{Exp}(A)P_{Th}(Bfollowing A) - P_{Exp}(Bfollowing A)P_{Th}(A) + P_{Exp}(A)P_{Exp}($

Probability Error of Mutually Exclusive Events

If A and B are Mutually Exclusive Events for the same experiment then;

Probability Error; P_E (A or B) $= P_E(A) + P_E(B)$

$= P_{Th}(A) - P_{Exp}(A) + P_{Th}(B) - P_{Exp}(B)$

$= [P_{Th}(A) + P_{Th}(A)] - [P_{Exp}(B) + P_{Exp}(B)]$

$P_E(A$ and $B) = 2[P_{Th}(A) - P_{Exp}(B)]$

For mutually exclusive events the sum of the probability error adds up to zero as the example on the coin shows.

Probability Error of Mutually Inclusive Events

If A and B are mutually inclusive events then;P_E (A and B) = $P_E(A)$ + $P_E(B)$ - $P_E(A)P_E(B)$

$$=2[P_{Th}(A) - P_{Exp}(B)] - [P_{Th}(A)P_{Th}(B) - P_{Exp}(A)P_{Th}(B) - P_{Exp}(B)P_{Th}(A) + P_{Exp}(A)P_{Exp}(B)]$$

$$=2P_{Th}(A) - 2P_{Exp}(B) - P_{Th}(A)P_{Th}(B) + P_{Exp}(A)P_{Th}(B) + P_{Exp}(B)P_{Th}(A) - P_{Exp}(A)$$

$$P_{Exp}(B)$$

$$=2P_{Th}(A) - P_{Th}(A)P_{Th}(B) + P_{Exp}(B)P_{Th}(A)$$

$$- 2P_{Exp}(B) + P_{Exp}(B)P_{Th}(A) - P_{Exp}(A)$$

$$P_{Exp}(B)$$

$$P_E(A \text{ and } B) = P_{Th}(A)[2 - P_{Th}(B) + P_{Exp}(B)]$$

$$-P_{Exp}(B)[2 - P_{Th}(A) + P_{Exp}(A)]$$

Probability Errors of Conditional Probability

The probability error of A given B is;

$P_E(A/B) = P_E(A \text{ and } B)/P_EB)$, $P_E(B)$ must not be zero

$$= P_{Th}(A)[2-P_{Th}(B)+ P_{Exp}(B)] -P_{Exp}(B)[2-P_{Th}(A)+P_{Exp}(A)]/P_{Th}(B)-P_{Exp}(B)$$

Binomial Probability Errors Theorem

If an experiment is a binomial experiment with two outcomes A and B and
$P_E(A)=P_{Th}(A) - P_{Exp}(A) = q$ then
$P_E(B)=P_{Th}(B) - P_{Exp}(B) = 1- q$,

and probability error of event E can be expressed as;

$$P_E(E) = {}_nC_rP_E(A)^{n-r}P_E(B)^r$$

$$P_E(E) = \frac{n!}{r!(n-r)!}q^{n-r}(1-q)^r,$$

where B occurs exactly r times in n

repetitions.

Expectation of the Number of Errors in a Trial

Expectation

The expectation of the number of errors in a trial = the theoretical probability error multiplied by total number of given trials at n. This can be defined as the expectation.

$$i.e. \ EP_E = (P_{Th} - \bar{P}m)[n(T)]$$

Experimental Expectation

The experimental expectation of the number of errors in a trial = the Mean Experimental probability error multiplied by total number of given trials at n. This can be defined as the experimental expectation.

$$\text{i.e. } EXP_E = (\overline{P}_n - \overline{P}_m)[n(T)]$$

Mean Deviation of Probability

The mean deviation can be given as; the sum of the mean probabilities minus the probabilities divided by the number sets.

But mean probability minus specific probabilities is the probability error. So mean deviation probability is equal to sum of the probability errors divided by the number of sets.

$P_{ER} = \dfrac{1}{S}\sum(P_{Th} - P_m)$	RANDOM
$P_{xn} = \dfrac{1}{S}\sum(P_{Th} - P_m)$	RANDOM

Standard Deviation of Probability

The standard deviation can be given ;

$\sqrt{\sum(P_{Th} - P_m)^2/s}$	RANDOM

$\sqrt{\sum (P_{Th} - \bar{P}_m)^2}_{/S}$	RANDOM

Part 2

The Relativity of Mutually Attractive/Repulsive Objects

chapter1

The Net Force of Mutually Attracting Objects

For forces which are attracting, the forces add up. If the forces F_1 and F_2 are attracting each other. The net Force is $F_n = F_1 + F_2$ and not $F_1 - F_2$.

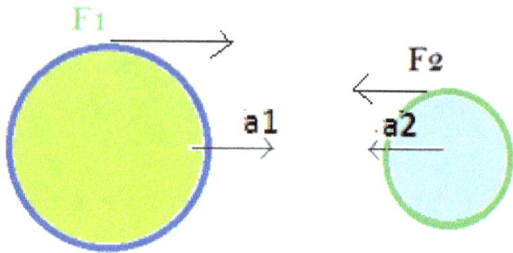

Suppose we have two objects attracting each other as shown, if the green object is kept still or fixed and the blue is left mobile. The force F_2 will have a reaction force on the blue object in the same direction as F_1 but equal to F_2. The resultant force, therefore, will be the sum of the two forces i.e.

$$F_n = F_1 + F_2$$

The Net Force of Mutually Repulsive Objects

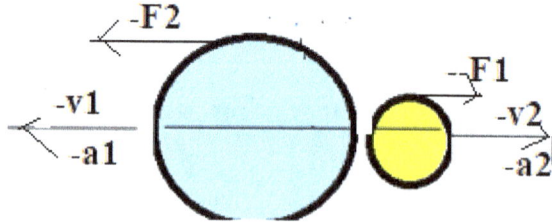

Similarly, for repulsive forces the Fnet will be

the sum of the forces but negative in value. The direction of attractive forces can be taken as positive so that the repulsive force is negative. Net force, therefore, will be $-F_1 - F_2$,

$$F_n = - (F_1 + F_2)$$

The Double Cartesian of Attraction/Repulsion

The formulae for net forces of attraction and repulsion show that they can be represented on a double Cartesian axis which face each other. The positive quadrants face each other and the negative quadrants face away from each other.

NEGATIVE REGION	POSITVE REGION	NEGATIVE REGION
negative	positive +VE	NEGATIVE

The middle area represents attraction and the negative areas represent repulsion. The resultant force can then be represented by a single Cartesian axes according to the usual rules.

The Correct Resolution of Forces in a 'Tug-Of-War' Game

The net force formulae also show that the correct resolution of forces in a tug-of-war is not $F1 - F2$ but $F1 + F2$ i.e. the forces add up. Suppose Group 1 is pulling with force of -F1 and group 2 is pulling with force --F2 then the rope of the tug-of-war has reaction forces F1 and F2 in it ie $F1 + F2$. The rope offers tension equal to this force. So both groups experience a net force of $F1 + F2$ at the point of contact. This is the force they feel in their grips not F2 or F1. This reaction force is balanced by the pulling forces –F1 and –F2 ie $^- (F1 + F2) = F1 + F2$. So the system as a whole has zero net force and no group moves until one side slackens. The same reasoning applies to tug-of-war of vehicles too.

Moderated Net Acceleration of Mutually Attracting Bodies

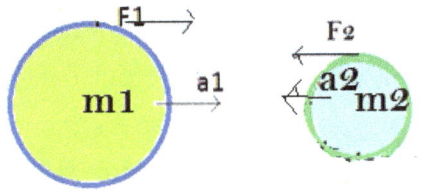

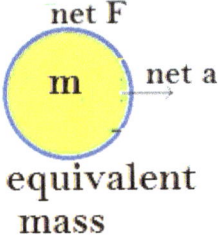

equivalent mass

Suppose we have two objects attracting each other as shown. The effect of F_1 will be to make object m_2 move with acceleration a_1, and the effect of F_2 will be to make the object m_1 move too with acceleration a_2. These movements will be in the opposite direction of each other. The accelerations a_1 and a_2 will be determined by the amount of inertia or masses of the two objects.

The movements of the two masses can be simplified into one movement, a net movement. We can consider the whole movement as that of one object with net mass, m_n , of the sum of the two masses i.e. $m_n = m_1 + m_2$. This implies that the net force, F_n , on m_n will be equivalent to the sum of the forces F_1 and F_2 acting on m_n i.e. $F_n = F_1 + F_2$.

The net acceleration of m_n, however, will be affected by the masses of the two objects. It will not simply be the sum of the two accelerations of the two objects i.e.

$$\boxed{a_n} \neq \boxed{a_1 + a_2}$$

as I previously stated. Instead the net acceleration will be moderated by the masses. How it is affected by the masses has been mind boggling until now. The following derivation shows how the net mass affects the net acceleration; $F_n = F_1 + F_2,$

$F_n = m_1a_1 + m_2a_2$, but $m_1a_1 + m_2a_2$ is not equal to $(m_1 + m_2)(a_2 + a_1)$ because it does not mathematically stand. So a_n cannot be equal to $a_1 + a_2$. The modified net acceleration will also be less than $a_1 + a_2$ since when $(m_1 + m_2)(a_2 + a_1)$, is expanded gives $m_1a_1 + m_2a_2 + m_1a_2 + m_2a_1$ which must be bigger than or equal to

$(m_1 + m_2)a_n$ at the least. Moreover, by the conservation of energy, we wouldn't expect the net acceleration to be greater than its components unless extra energy is applied to the system. So net acceleration cannot be greater than $a_1 + a_2$.

We saw that m_n is the equivalent mass of the two masses in the system. So $m_n = m_1 + m_2$ and $F_n = F_1 + F_2$. Now there must be a certain acceleration, a, such that $F_n = m_na$, which gives $m_na = (m_1 + m_2)a$,

but $F_n = F_1 + F_2$, which is $F_n = m_1a_1 + m_2a_2$,

$m_na = m_1a_1 + m_2a_2$, then equating

$(m_1 + m_2)a = m_1a_1 + m_2a_2$

And hence $$a = \frac{m_1a_1 + m_2a_2}{m_1 + m_2}$$

So this is the net acceleration moderated by the masses. We can therefore call it the moderated net acceleration $\mathbf{a_m}$. It is not equal to the net acceleration i.e. $a_n \neq a_m$.

$$\text{So } \mathbf{a} = \frac{m_1 a_1 + m_2 a_2}{m_1 + m_2} \text{ becomes}$$

$$\boxed{\mathbf{a_m} = \frac{m_1 a_1 + m_2 a_2}{m_1 + m_2}}$$

Moderated Net Acceleration of Mutually Repulsing Bodies

For repelling objects, the repulsive forces will set up reaction forces R1 and R2 on the opposite objects. The net force will cause accelerations $-\mathbf{a_1}$ and $-\mathbf{a_2}$ on the two objects.

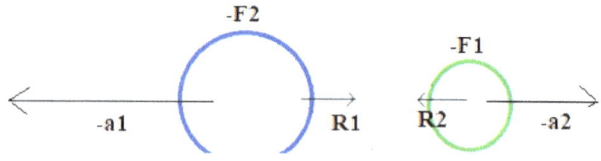

The net acceleration will therefore be;

$$a_m = \frac{-(m_1 a_1 + m_2 a_2)}{m_1 + m_2}$$

This is the moderated net acceleration that the object with the smallest force in the system will move.

CHAPTER 2

Galileo was Wrong: Masses do not Fall at the Same Rate

Net Force due to Gravity, F_n

Newton said that all objects attract each other i.e. all objects have gravity. When placed near each other, the objects will therefore fall into each other. If object 1 has force 1 and object 2 has gravity force 2 then the net force will be $F_1 + F_2$. The net force therefore between the two is greater than their individual gravitational forces.

So $\boxed{F_n = F_1 + F_2}$

This Fnet will be different for two objects of different masses brought near the same object because of differences in their forces of gravity.

So the object-earth system with greater net force will have higher rate of falling than the object-earth system with less net force.

The Complete Definition of Weight

The force the earth exerts on the object near the surface is called the weight. So $F_1 = W_1$ and the F_2 can be called the weight of the earth on 'planet object' so that $F_2 = W_2$. So

$$\boxed{F_n = W_1 + W_2}$$

But w =mg so we get;

$$F_n = m_1 g_2 + m_2 g_1$$

or $\boxed{F_n = m_e g_o + m_o g_e}$

for objects near each other.

But the force between two objects is also given by $F_g = \dfrac{G \dfrac{M_1 M_2}{r^2}}{}$

so $\boxed{m_e g_o + m_o g_e} = G\dfrac{M_e M_o}{r^2}$ or

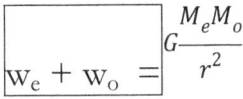

$$W_e + W_o = G\frac{M_e M_o}{r^2}$$

where W_e and W_o can be called the particular weights of the earth and object respectively. The definition of weight is given as;

$M_o g_e$ and is equated to $G\frac{M_e M_o}{r^2}$. But this definition ignores force of the object on the earth. The complete definition of weight, therefore, should include the force of the object on the earth or planet. The complete definition should be the sum of the two forces involved in the 'tug-of-war'. It should be; $m_e g_o + m_o g_e = G\frac{M_e M_o}{r^2}$

i.e. $$W = m_e g_o + m_o g_e$$

OR

$$W = W_e + W_o$$

We must be talking about the net weight therefore and not just weight. All weights we measure are net weights. The paradigm must shift.

The Weights of the Planets

When we measure weight using a spring balance we are actually measuring this net force or net weight. The particular weights of the object alone or earth alone can never be found i.e. w_e and w_o. We can only approximate W_O of astronomically small objects since their masses are very tiny in comparison to the earth's mass. Therefore, the object is accelerated more than the earth or planet. This gives rise to a bigger weight for the object than the earth or planet. So the planet's particular weight on the object can be ignored.

The particular weight of the planet or earth on the astronomically small object is very small almost zero since the pull of gravity of the object acting on the planet or earth is very small and hence makes the planet or earth move very slowly or not at all. Thus the small acceleration gives rise to a very small weight. This weight cannot, therefore, be approximated. Moreover, it is not of practical significance to the scientist in relation to the weight of astronomically small or everyday objects. It seems the concept of weight is practically useless compared to the concept of mass. More comparisons can be done using mass than weight. Mass is more practical than weight.

When we measure the weight of the object we are simultaneously finding the weight of the earth/planet on that object. It is like the faces of a coin or the poles of a magnet; they can never be separated to stand on their own. In short, weight is reciprocal.

So the same net weight of the object on the earth/planet is also the net weight of the earth/planet on that object i.e. $m_e g_o + m_o g_e = W$.

It is not the weight of the Earth/planet only or weight of the object only but of both.

For instance, for an isolated system of an object of mass 60kg and earth then the weight of the object on earth is approximately 60 x 10 = 600N and the weight of the earth on the object is also 600N. This 600N is the weight of the object plus the weight of the earth.

The weight of the object alone is not 600N.

It is less than 600N and the weight of the earth on the object is not 600N but less almost zero.

Since it depends on the net g, it is possible for the complete weight of the earth or any other astronomical body to be small or even zero. Since the pull of the object on the earth/planet is small, the earth/planet accelerates slightly or remains stationary. The force/weight, therefore, will also be small or zero in spite of the large mass of the earth/planet.

Approximating g of Small Objects

This implies that we can get an approximate of the g of the object since the mass of the earth is known. We can ignore the weight of the planet on the object and use the formula $W = mg$. So for a 60kg object whose weight is approximately 600 which is equal to approximately $m_e g_o$. Since the mass of the earth has factor 10^{24}kg, we get g_o in factor of 10^{-22} N/kg or 10^{-22}m/s^2. For a 20,000 tonnes ship i.e. 20,000,000kg the weight is approximately 200,000,000N . So the g will be approximately in the factor of 24^{-16}N/kg or 24^{-16}m/s^2. We can readily see that even for some of the largest manmade structures the g is still very small. The difference between the g of a 60kg object and that of a ship would be in a factor of 24^{-16}m/s^2. This difference would be hard to observe for the two objects.

Moderated Net Acceleration due to Gravity, g_n

In the previous editions I had stated that an object m_1 coming from outside the other object's field, will accelerate towards m_2 with acceleration of a_1, and m_2 will accelerate towards m_1 with acceleration of a_2 and the net g will be the sum of the two gs i.e. $g_n = a_1 + a_2$.

But it has come to my attention that this cannot be since the accelerations of each object will be affected by the masses.

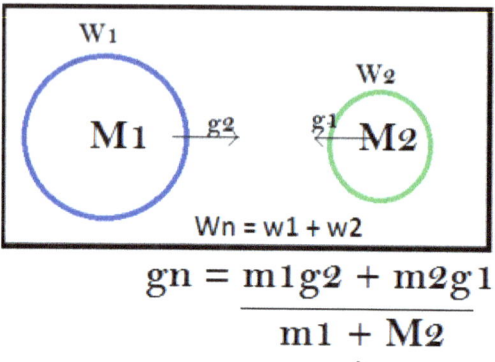

$$gn = \frac{m1g2 + m2g1}{m1 + M2}$$

If an object of mass m_1 with g of g_1 is brought near the earth or planet with mass m_2 and g of g_2 then their masses merge together to give m i.e. $m_1 + m_2 = m$.

But their net g, g_n, will not be the sum of their gs i.e. $g_1 + g_2$ as we saw previously with mutually attracting objects. This is because the new weight of m, W_n, is not the sum of their weight, as we previously saw i.e. $W_1 + W_2 \neq W_n$ or $m_1g_2 + m_2g_1 \neq (m_1 + m_2)(g_2 + g_1)$. It does not mathematically stand. Instead, the net g will be a moderated g_m; so $(m_1 + m_2)g_m = m_1g_2 + m_2g_1$

$$g_m = \frac{m_o g_e + m_e g_o}{m_o + m_e}$$

$$\text{or } \boxed{g_m = \frac{1}{M} (W_o + W_e)}$$

where $m = m_1 + m_2$ or mass of object and earth, W_o is $m_1 g_2$ or particular weight of object and W_e is $m_2 g_1$ or particular weight of the earth.

So the new g changes and must be smaller than the sum of their old gs. This is also the net g that an object-earth system will experience not $g_1 + g_2$ as I had previously stated.

This moderated net g experienced by an object when falling on earth or any other planet is also the g experienced by the earth or planet. We must no longer talk about a specific g of the earth or object. The only g we can talk about is the g net of each earth/planet-object system. The gs are specific to the masses and not general.

Objects do not Fall at the Same Rate

In defending Galileo's assertion that masses fall at the same rate in spite of differences in their masses, someone claimed that somehow, mysteriously, the masses and forces in gravity field balance each other in such a way that the acceleration due to gravity is always a constant. Someone else claimed that the bigger mass cancels out the expected higher acceleration by moderating the force since $g = F/m$ so that the g remains the same.

How that moderation or cancelling out is done has been mind boggling and a mystery until now.

This relationship $g_m = \dfrac{m_1 g_2 + m_2 g_1}{m_1 + m_2}$ shows how the moderation is done or how the g is affected. It proves Galileo was wrong and the mysterious moderation or cancelling out wrong. The moderation or cancelling is not in such a way as to make the g constant.

The relationship $g_m = \dfrac{m_1 g_2 + m_2 g_1}{m_1 + m_2}$

which we can rewrite as $g_m = \dfrac{m_o g_e + m_e g_o}{m_o + m_e}$ where m_o is mass of object, g_o is the g of the object , m_e is the mass of the earth and g_e is g of the earth, shows that moderated net g is not constant. It is moderated by the mass through the relationship $g_m = \dfrac{m_o g_e + m_e g_o}{m_o + m_e}$.

All the objects we deal with are astronomically small and hence their g s are very small. When we replace this small mass and small g in the formula, the numerator in the formula $g_m = \dfrac{m_o g_e + m_e g_o}{m_o + m_e}$ gets smaller compared to the denominator. The denominator in the formula is minimally affected since we are adding a small mass to a huge mass. The numerator, however, is affected considerably and gets smaller since we are multiplying the mass or g of the earth with a very small number of g_o or m_o. The mass and g of the earth remains the same throughout but the objects have different masses and gs. So m_o and g_o varies but m_e and g_e does not.

It also seems that m_o is directly proportional to g_o so that as g_o tends to zero m_o also tends to zero though it lags behind since it is considerably larger than g_o.

So g_n tends to zero as g_o and m_o tends to zero. But for a massive object like the moon or Jupiter, with more g, the numerator gets considerably bigger since we multiply the big mass or g of the earth by a big g_o or m_o. But since we are just adding the two masses on the denominator, the denominator will be smaller than the numerator. So the g_m gets big. It also seems that m_o is directly proportional to g_o so that as g_o tends to infinite m_o also tends to infinite.

So g_m tends to infinite as g_o and m_o tends to infinite.

Galileo was Wrong

This clearly shows that Galileo was wrong; objects do not fall at the same rate. The g_m of the bigger object is bigger than the g_m of the smaller object. In short, heavier objects fall faster than less heavy objects.

But because we deal with mostly tiny masses (which I will call, astronomically small masses since they are considerably smaller than the earth no matter how big they seem to us), the difference between their masses are negligible and hence their net gs differ slightly as to give approximately the same net g when plugged into the formula; $g_m = \dfrac{m_o g_e + m_e g_o}{m_o + m_e}$.

Galileo Galilei was wrong. His famous experiment was an oversimplification which led to error because it was not accurate enough.

The Flaws in Galileo's Experiment

Galileo's experiment on the Leaning Tower of Pisa had four flaws;

 i) The earth-object system was to be isolated for each object so that the gravity of one does not interfere with the gravity of the other. The gravities of Galileo's feather and stone acted as that of one mass and force in relation to the earth. They therefore fell at the same rate. If they were separated and shielded from each other, Galileo could have observed that they fell at different rates.

iii) The difference between the two masses was not sufficiently great to show the difference in the rate of fallings. Because Galileo dropped very tiny masses with very small g in comparison to the earth, the net g for each dropping differed very slightly as to be imperceptible. If he had done the same experiment with a very tiny object like an atom and another massive object like the moon, the moon would have fallen faster than the atom i.e. g_m moon and earth system will be greater than g_m for an atom and earth system. The moon has g $1/6$ th that of the earth and an atom must have g significantly smaller than that of the moon. So g net of moon-earth system will be greater than that of atom-earth system. In short, the moon will fall faster to the earth than the atom whose g net is less.

Contradictions if Galileo was Right

But someone may object. They may say that the mass or inertia of the moon or massive object may actually slow down the rate of falling hence cancelling out the effect of the extra mass and making the rates of falling of all objects equal. But in whose favour will the cancelling out be? Moon or earth? If we maintain the status quo and say all objects fall on a planet with a constant g of that planet then we would expect the earth to fall into moon with the g of moon which is less than that of the earth. The observer on the earth will therefore see the moon as falling into the earth slower than usual i.e. at the g of the moon $1.46m/s^2$. We cannot expect the moon to fall at the g of the earth because we are considering the earth as the object.

Or we can consider the moon as the object falling on the earth with the g of the earth since all objects fall on the earth with a constant g of $9.81m/s^2$. We, therefore, have a contradiction and we are stuck in a quandary. So this cannot be. We cannot maintain the status quo. The moon cannot fall at g of the earth and at its own g at the same time. It must fall at a certain g which must be some combination of the two. Galileo was therefore wrong; objects do not fall at the same rate.

Moreover, we have already seen that the relationship $g_m = \dfrac{m_o g_e + m_e g_o}{m_o + m_e}$ shows that the heavier object will fall faster than the less heavy object on earth or a planet.

So the moon will fall on earth at approximately; $g_m = \dfrac{m_o g_e + m_e g_o}{m_o + m_e}$,

$$g_m = \frac{10^{22}(10) + 10^{24}(1.64)}{10^{22} + 10^{24}}$$

We usually deal with astronomically small objects which are very close in g. So Galileo's stone and feather were relatively very close in mass even though to us they seem to have big difference in mass. The difference in masses was not big enough to give a perceptible difference in the falling rates.

Nevertheless, we can restate Galileo's assertion as; objects with relatively small difference in their masses and hence in their gs fall approximately at the same rate whether those objects are astronomical or microscopic in size.

If the difference in masses between them is considerably big then the rates of falling will also be considerably big and perceptible to human beings.

Factors that Determine Size of g_O

It seems that the accelerations due to gravity of objects and planets are directly proportional to the masses and density of the objects or planets. For instance g of earth is $9.81 m/s^2$, for the moon it is $1.64 m/s^2$.

$$\text{So} \quad g_o = k \rho m_o , \quad g_o = k^{\frac{m_o}{V}} m_o$$

$g_o = k m_o^2/v$, but $V = \frac{4}{3}\pi r^3$,

$$\text{so} \quad \boxed{g_o = k \, 3m_o^2/4\pi r^3}$$

k is constant of proportionality which can be called constant of g or constant of weight. M_o should not be confused with mass of the object for which the weight is being found in relation to that planet.

Weight can also be found on any object and not only on a planet. So the notation of g_o and m_o is more appropriate where O stands for object. All this implies that weight must be actually lower than what we measure it.

This formula shows that the magnitude of g in the object varies directly as the square of the mass and inversely as the cube of the radius from the center of the earth i.e.

$$g_o = k_w m_o^2 / r^3$$

k_w being a new constant of the weight.

Measurement of g

The measurement of g by pendulum is affected by this fact. The mass of the bob and it's gravity is tiny to show the real g of the system. More-over everything will be affected by the earth-universe system so the net g is the result of all the gravity of the universe and the earth.

Measurement of g and New Formula

The measurement of g by pendulum is affected by this fact. The mass of the bob and it's gravity is tiny to show the real g of the system. More-over everything will be affected by the earth-universe system so the net g is the result of all the gravity of the universe and the earth.

The period of the pendulum will be affected by the mass. It is no longer independent of the mass. If we replace $g_o = k\rho m_o,$ into

$T = 2^\pi\sqrt{\dfrac{l}{g}}$ we get $T = 2^\pi\sqrt{\dfrac{l}{k\rho m}}$ but $\rho = \dfrac{M}{V}$, So

$g = km^2/V$

$T = 2^\pi\sqrt{\dfrac{Vl}{KM^2}}$ but for a spherical bob $V = 4/3$

πr^3, $T = 2^\pi\sqrt{\dfrac{4\pi r^3 l}{3KM^2}}$, $T = 2^{\pi(2\pi^2)}\sqrt{\dfrac{r^3 l}{3KM^2}}$, $T = 4$

$\pi^3\sqrt{\dfrac{r^3 l}{3KM^2}}$, $T = \dfrac{4\pi^2}{9k^2}\sqrt{\dfrac{r^3 l}{M^2}}$, $T = \dfrac{4\pi^2\sqrt{lr^3}}{9k^2\ M}$,

$T = \dfrac{4\pi^2\sqrt{lr^3}}{9k^2\ M}$, $\left(\dfrac{2\pi}{3K}\right)^2$ can be called the Pendulum

Constant P.

$$\text{So } \boxed{T = P}\,\dfrac{\sqrt{lr^3}}{M}$$

This formula shows that the period varies inversely as the mass. We have been teaching that mass is not a factor, but when the mass is huge or when using a very sensitive pendulum, it is a factor.

The formula also shows that the period varies directly as the radius of the bob. This is just a consequence of the inverse square law of gravitational force.

Galileo's hunch, which he was actually testing

in the experiment at the tower of Pisa that the rate of falling is directly proportional to the density of the object was actually correct but the rate also depends on the mass. For the same substance with the same density the more massive object will fall faster than the less massive one. This is because the more massive object has more electrons and protons which cause more gravity force than for the less massive object.

Pendulum in a Vacuum

An experiment can also be done on earth using pendulums in a vacuum. Since the slowing down of a pendulum is due to air drag, a pendulum in a vacuum will swing for ever or very long time until stopped by the small friction force between the particles in the string as it bends due to the swings.

Two pendulums can be constructed of the same length but different masses. The pendulums can be dropped from the same height. Since the bigger mass should drop faster, it will eventually overtake the smaller mass when it reaches the bottom. This overtaking may not be perceptible initially, but overtime the bigger g or acceleration of the bigger mass will make exponential increments in the difference of the two pendulums at the bottom. This will prove that the g is not constant and that the period is affected by the mass.

Rate of Falling Formula

From the equations of motion, the time of falling for objects into the earth will be;

$$t_o = = \frac{V_e\sqrt{g_e}}{g_e\sqrt{g_o}}$$

t_o , v_o being time for object and velocity of object. The time for earth falling into object will be;

$$t_e = = \frac{V_o\sqrt{g_o}}{g_o\sqrt{g_e}}$$

We do not have the sense of us moving into the opposite direction. To us only the moon or falling object will seem to be moving/falling.

For an observer on the earth or planet, objects therefore do not fall at the same rate. It is like jumping up to catch a falling ball or object; the object is moving towards you at about 9.81 m/s^2 and you are also accelerating towards the object. You therefore meet the object earlier than if one did not jump. The net g of the jumping person and the falling object is therefore greater than 9.81m/s^2. The g of the ball to the person who did not jump was the normal g of 9.81m/s^2. To that observer objects fall at different rates. We have been wrong for more than 400 years.

ii) The distance from which the objects were dropped was also not sufficiently great to show the difference in the rate of fallings. If two smaller objects smaller than the earth by a factor of a billion and hence their gravities were of a factor of let's say 10^{-6} but differed slightly (like Galileo's feather and stone or Scot's feather and hammer), then by; y = -0.5gt^2, they would fall a distance of ;

6 x 10^{-5}m in 120 sec i.e. 2 minutes, and

6 x 10^{-4}m in 1200 sec or 20 minutes, and

6 x 10^{-3}m i.e. 0.006m or 0.6cm or 6mm in 200 minutes or 3.3 hours.

So Galileo would have required a very high height or very big difference in the masses of the objects to detect the difference in the falling rates between them. A height which would give a difference of 0.6cm or 0.006m would require a height of;

$y = 0.5(9.81)(12000)^2 = 706,320,000$m
or 706,320km.

So just to observe a difference of 6mm in the falling rates for small objects would have required a height or ramp more than 700 thousand km up into space, and a waiting time of more than 3 hours to see just one falling object reach the ground.

In short, there was a slight difference in the rate of falling of the feather and stone but Galileo could not easily observe it. With the speed at which the object hit the ground, Galileo could not have possibly differentiated the rates with his naked eyes. But to his credit in his rigorous experiment, he managed to slow the speed using ramps. But they were not sufficiently high and the objects did not sufficiently differ in mass. They worked correctly in determining that the rate is exponential in form but it is not the same for all objects. But one can claim that proves that g is approximately constant. But we have seen that if the difference in masses between the two objects being compared is great and the experiments done one at a time then their rate of fallings will not be constant.

So it's not the height of dropping which is a crucial factor but the difference in the masses being compared. The smaller the difference the less the difference in the rate of fallings and the more the difference the more apparent and great will be the difference of the rates of fallings of the two objects.

iv)Galileo did not know Newton's law of Universal Gravitation. He did not know that the earth was also falling into the feather and the stone. He did not take into consideration the movement of the earth into the opposite direction.

Newton, too, could have proved that Galileo was wrong. But I guess he couldn't fathom challenging the great master. Newton could not see the implications of his own law of Universal Gravitational in relation to rate of falling bodies.

Conservation of Position

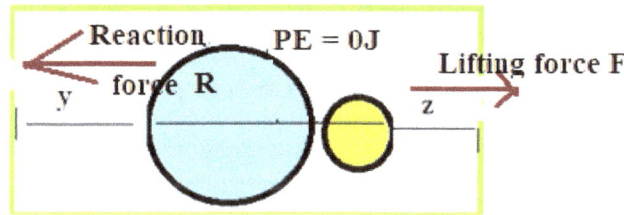

For objects in contact with each other and in the same field, when the object is lifted up from the earth, the earth also moves backward due to the reaction force by Newton's third law. When the objects falls back, the earth also moves towards the object due to the gravitational force. The two objects will meet again on the same position. So for an isolated system position of the objects is conserved.

The Reciprocity of Gravitational Potential Energy

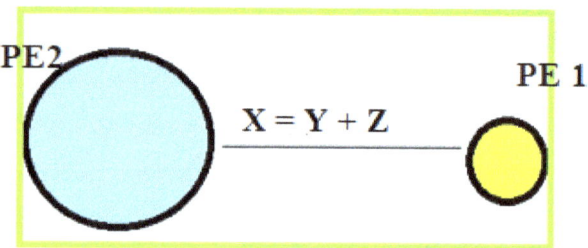

The PE of the two objects in relation to each other is zero when the objects are in contact. When the smaller object is lifted up, it gains potential energy PE1 in relation to the bigger object and distance Z , and PE1 = $M_o g_e y$. But $M_o g_e$ is weight W_e.

The bigger object gets PE 2 in relation to smaller object and distance Y and PE2 = $M_e g_o y$. but $M_e g_o$ is weight W_o , PE1 = $M_o g_e$.

If one object is considered stationary and the other as moving then PE = Mgy = Wy. But W = $m_e g_o$ + $m_o g_e$,

So $\boxed{PE = (m_e g_o + m_o g_e)y}$
$$\boxed{PE = W_e y + W_o y}$$
$$\boxed{PE = Wy}$$

This PE is also the PE of the earth in relation to the earth i.e. PE is reciprocal just as weight.

When the objects fall back into each other, the PE is again back to zero. So the energy of the system is conserved and the position of the system is conserved i.e. the earth object system moves back to the original position.

Anti-gravity, Anti-weight and Anti-g

From our section in chapter 3 on repulsion of bodies, if we have two repelling astronomical bodies then we can talk about gravity. The force of repulsion could be called anti-gravity, the net acceleration between them could be called anti-g and the force between them could be called anti-weight.

The Anti-Gravity Potential Energy

If two objects are repulsing then we can talk about anti-gravity Potential energy. The formula will be the same as for the

Gravitational PE except that it will be
negative i.e.
$$APE = -(m_e g_o + m_o g_e)y$$
$$APE = -(W_e y + W_o y)$$
$$APE = -Wy$$

This PE is also the PE of the earth in relation
to the earth i.e. PE is reciprocal just as weight.

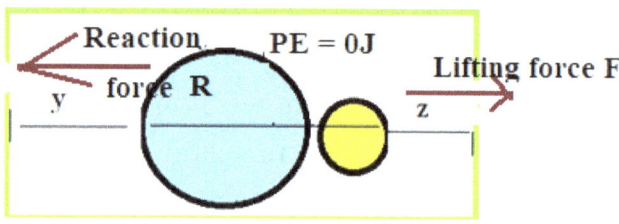

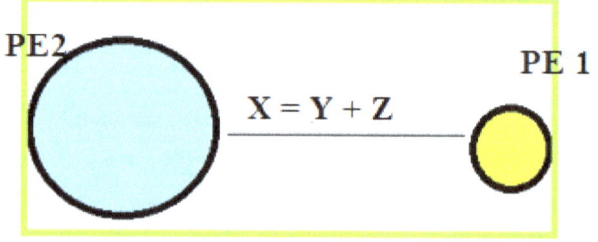

CHAPTER 3

RELATIVITY OF ACCELERATION

The Galilean Mutually Attracting Bodies

For two bodies in space each object is falling into another's gravity field at that object's g and the other object is also falling into that object with that object's g, they will meet faster and some way through the journey. The relative acceleration of one body from the observer on the opposite body, and vice versa, will be greater than if only one body has gravity and the other did not.

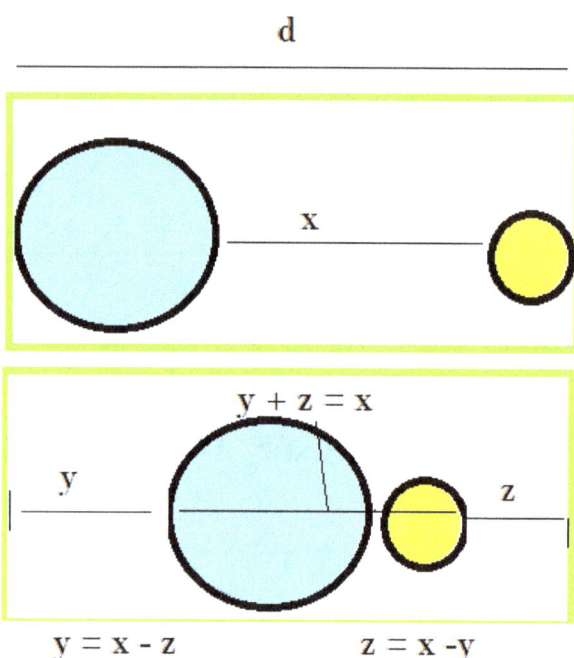

The blue object will fall into z by y meters while the yellow will move z meters. Instead of falling through the length x, the smaller object will seem to fall through $x - y$ to an observer on the yellow object, and instead of falling through y, the bigger object will seem to fall through $x - z$.

Also instead of taking the time t_x to meet, the objects will take t_{x-y} and t_{x-z} to meet. So instead of the acceleration being $\mathbf{a_{xe}} = \mathbf{V_1/t_x}$,

it will be $\mathbf{a_{x\text{-}z}} = \dfrac{V_1}{t_{x\text{-}z}}$. Since V is the same but $t_x > t_{x-z}$, $\mathbf{a_{x\text{-}z}}$ is greater than $\mathbf{a_{xe}}$. Also the acceleration of object $\mathbf{a_{xo}}$ will be different from $\mathbf{a_{x\text{-}y}}$.

It is like jumping up to catch a falling ball or object; the object is moving towards you at about 9.81 m/s^2 and you are also accelerating towards the object. You therefore meet the object earlier than if one did not jump. The net g of the jumping person and the falling object is therefore greater than 9.81m/s^2 for an observer in space who does not see the ground as a reference point of the two motions. To that observer objects fall at different rates.

So for an observer on the yellow object, oblivious to his movement towards the blue object, will see the blue object fall faster than normal and vice-versa.

The Galilean Mutually Repelling Bodies

If the two objects are repelling then the relative velocity noticed by an observer on one object at lower speeds will be the sum of the two velocities as they recede from each other i.e. $\mathbf{a} = \dfrac{-(v1 + v2)}{t}$ instead of $\mathbf{a} = = \dfrac{-v1}{t}$. Since the time is the same in both cases but the observer notices one velocity as sum of the two velocities the net acceleration is bigger than expected.

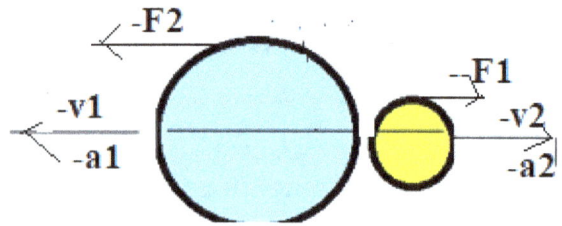

CHAPTER 4

ELECTROMAGNETIC WAVES AND ENERGY

Fields are Like Elastic Bands

Experience with repulsion of magnets shows that fields can be considered elastic. When opposite fields are compressed together, they rebound. So fields are like a series of concentric stretched elastic bands connected at the center. Each band contains a certain amount of energy given by Planck's Formula; $E_n = nhf$.

The tension of the field is like the tension of the elastic band and so is the magnetic force. The force towards the center is like the force of the rebounding elastic band and so is the electric force.

Therefore each band has its own electrostatic elastic potential energy. In short, the electric potential energy is the electrostatic elastic potential energy

i.e. $EEP_f = \dfrac{1}{2} Kx$, were x is the distance of the field from the charge carrier and k is the constant of elasticity. So at each point EEP =

U, $\dfrac{1}{2}$ Kx = QV,

$V = \sqrt{\dfrac{Kx}{2Q}}$, k being the constant of elasticity or spring constant.

The Electromagnetic Waves and Energy

The electric fields can be considered like elastic bands which have been stretched from the center of the electron. atom or object. Each position in a field will correspond to a single stretched elastic band a certain distance and direction from the centre. So each band or position has elastic potential energy.

The disturbance of the fields or bands, stretching them or compressing them, either increases the elastic potential energy or decreases it. This band disturbs the next band next to it, disturbing its elastic potential energy too. This band does the same to the next band. A wave therefore is created. This is the electromagnetic wave.

The field/band is stationary, it is an electric field. When it moves it becomes a magnetic field.

The Energy of the Fields

The energy as a result of the disturbance does not travel but oscillates within that band. When the field is stationary the energy is elastic electrical potential energy and the field is purely electrical field. When the field moves, this energy is converted to electric KE and the field is a magnetic field.

The disturbance in the energy reaches a certain point where it can be observed as light, heat, and gamma rays etc i.e. as EM waves. The EM waves do not travel to reach us; it is the disturbance travelling at the speed of light that reaches us just as the water particles do not travel in a water wave. What is disturbed in our eyes or ears when we sense light or heat are our electric fields in the eyes or skin.

Implications on KE Theory

To lose or gain energy is to have the fields

disturbed around the particle. The particles are in constant motion because the fields are always losing/gaining energy/having their fields disturbed to a less or greater extent constantly.

Zeroth [Z] Waves and Eternal [E] Waves

The frequency of the disturbance of the bands produces each type of wave. Since the particles are always vibrating, they are always producing EM waves even when we can't detect those waves.

The waves from 0Hz to the radio waves which we can't detect can be called the Zeroth waves or infra- radio waves. All vibrating matter apart from producing sound waves also produces EM waves in form of infra- radio waves. It is therefore possible to detect sound without using a microphone but a receiver circuit tuned to an infra-radio wave frequency.

Since waves are produced when a particle is accelerating, a particle can accelerate indefinitely until it reaches the speed of light in one direction. For indefinite acceleration to occur the time must tend to zero and the change in velocity must also tend to zero.

When the particle reaches the speed of light then it will no longer accelerate and the EM waves will cease being produced since the particle is in constant speed. The EM waves from Gamma to zero can be called Eternal waves, E waves or ultra gamma rays.

The electromagnetic spectrum is therefore circular.

When the wavelength is zero and the amplitude infinitely long then we get the maximum frequency of EM waves.

When the wave length is very long and the frequency so low that the sine wave becomes a line then we get Z waves.

Energy, Frequency and Position

For the same amount of energy, the frequencies in each band will move from lower to higher, analogous to strumming strings of a guitar, since each band has its own length due to its distance from the centre. So the distance from the center correspond to the frequency and length multiplied by the frequency.

 For the same amount of energy, Frequency is inversely proportional to distance of field from the centre i.e.

F $=k/x$, k = Fx, the k is the energy since this occurs for the same amount of energy.

Energy = frequency $\times$ radius of field
$$E = fr$$

The Radius of the Photon

From $E = fr$, $r = E/f$. When the radius is constant, that is; when energy is fluctuating within one field, the radius' magnitude equates to the magnitude of the Plank Constant h i.e. nhf = fr

$$r = nh$$

This gives the radius of the specific

electromagnetic field or photon. This radius corresponds to the energy as already stated. So the Gamma ray field has the greatest radius and the radio waves field has the least radius. If the n corresponds to the electromagnet spectrum, then n for gamma radius will be n x 6.62 x 10^{-34}j.s =

3.97n x 10^{-34}m and that of the radio waves will be   6.62n x 10^{-34}m

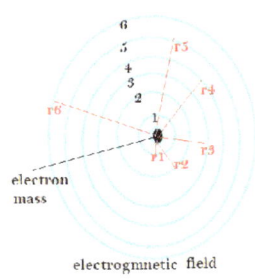

electrogmnetic field

Why Speed of Light is Constant

Einstein never bothered to find out the reason for his law of constancy of speed of light. I brave a conjecture;

Light and EM waves can be considered like particles with mass ejected when a force or impulse is applied on the fields. The amount of agitation is proportional to the frequency of the field just as in a guitar string the impulse on the string is proportional to the frequency and hence the note. In a guitar, we can talk about the impulse at that specific position of the string or area. So we can define another quantity which can be called impulse pressure, IP,

then $\qquad$ IP $= \dfrac{F}{At}$

and the units are N/sm^2,

for a guitar, this impulse is proportional to the frequency i.e. IP = kf, fk $= \dfrac{F}{At}$, f = number of waves per time, so number of waves $= k\dfrac{F}{A}$, number of waves is directly proportional to the pressure applied.

But the 'note' produced in a field is not a sound but an EM wave. This EM field will vibrate and eject the 'EM' waves/ particles out. If we consider these fields and their EM waves like particles with the smallest mass in the universe, then the slightest impulse pressure or force causes them to be ejected. And if the speed is constant, then in that instant-time they accelerate from zero to c and back again. From $a = \dfrac{V-u}{t}$, then $a = c/t$, we can call the a the acceleration of light, a_c and, t ,the instant time, t_i so, $a_c = c/t_i$

$c = a_c t_i$

the force that causes this acceleration can be called Force of light, F_L and the mass can be called the mass of light M_L. So that

$$F_L = M_L a_c$$

This force is the force in the Universe that causes EM waves. But just like a stretched string which can be strummed to a certain tension before it breaks, the fields can also be acted upon by a certain impulse before it breaks. $F_L/At_i = M_L a_c/At_i$,

$$\text{ie IP} = \dfrac{M_L a_c}{At_i}$$

This IP is very great because the small force is divided by very small area and instant time, which we can say is the smallest time possible. So at this impulse pressure/force the EM waves are released, but beyond this force the fields break apart creating two fields just as a guitar string can break into two strings.

So the EM waves travel at constant speed for this reason. Any slight increase of the force beyond F_L results in breakage of the fields.

The Fields Correspond to Each EM Wave

For the same amount of energy, the shortest field will produce the most frequency, and the longest field will produce the lowest frequencies. Since these frequencies correspond to types of EM waves, the levels of the fields correspond to EM waves. So we can have Z fields, radio fields, etc and so on up to gamma fields and E fields.

It is possible for one level of field to produce all these EM waves depending on the energy. So the longest field can produce the gamma rays if enough force disturbs it. This means these other fields will produce more than gamma rays i.e. they will produce Ultra-gamma waves or ultra—ultra gamma waves.

Redefining Matter

A piece of matter is like a solid core with a jelly mass around it. The core is the mass with its centre of mass and the jelly mass is the electric field.

Every object has the jelly mass/field around it even neutral objects. They have a thin layer of field due to polarization. This is equivalent to the Quantum Field though a field itself should not be described as a particle. It may behave like a particle but it is not a particle.

For a particle to interact with another particle, it must have an electric field. For a particle to have directions and hence interact effectively, it must have the centre of mass. A particle with-out an electric field but with mass only is not matter, and it can never interact with other matter.

These fields contain energy which correspond to the EM waves. The agitation of these fields causes the energy to oscillate from potential to KE and in between and then back.

81

This oscillation or its frequency can reach our eyes or skin and the brain detects that frequency as light or heat. The energy is not ejected out. It just oscillates or transforms.

These fields can interact resulting in electromagnetism and gravity.

The electric fields spread out in space. On this electric field are levels of energy which correspond to the electromagnetic waves.

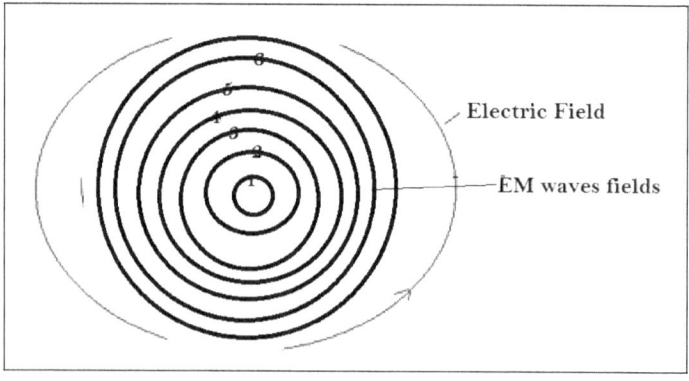

REFERENCES

1. Nelkon M, 1993, Principles of Physics, Longman, London.
2. Serway, R. and Jewett, J. 2004, 6th Ed, Physics for Scientists and Engineers, Thomas Brooks/Cole,
3. Engineers, Brooks/Cole, Boston.
4. Demana F, Waits B, Clemens S, 1994, 3RD Edition, Precalculus Mathematics; Addison –Wesley, Californa.
5. Various Internet Resources

www.ingramcontent.com/pod-product-compliance
Lightning Source LLC
Chambersburg PA
CBHW071028220526
45467CB00004B/1560